FROM INVISIBLE TO VALUABLE

Networking in the Age of AI

Turn Conversations into Clients,
Partnerships, and Opportunities

Doron Noyman

ISBN: 978-1-7361555-8-5

CONTENTS

PREFACE

Can AI just solve networking for me?

That was the question a young woman named Tracy asked me after a talk I gave on building professional relationships. Tracy had started a small business a few months earlier. She baked cakes, cupcakes, and catering orders out of her home kitchen. Her dream was to open her own bakery. Other entrepreneurs had told her that relationships would help her get there. Local business groups emphasized it constantly. Yet after months of effort, she still hadn't landed consistent clients. But she had also been thinking about AI. She'd watched tons of videos about it, read articles, and convinced herself that if she just learned the right prompts, technology would handle the hard parts of networking: awkward introductions, endless follow-ups that never led anywhere, and events that were a waste of time. It wasn't working—not even with AI. Conversations were still awkward. The connections still faded before she could capitalize on them. Now she was starting to wonder if networking worked at all. Her question about using AI to solve networking problems made me pause and think, not because it was naïve, but because it was exactly the right question for this moment. I had been hearing versions of it everywhere. *If AI can do so many things, can it solve networking, too?* The question comes up after presentations, during business meetings, and at networking events. Seeing how quickly AI is changing the way people work, professionals and small business owners are naturally curious: how can AI fit into something as personal as building relationships? I empathized with Tracy's frustration because I had spent decades in those same rooms, at the same networking events—business roundtables, Chamber of Commerce meetings, and industry conferences. As a past chair of the board of the Silicon Valley Chamber of Commerce, and through years of work alongside entrepreneurs and professionals across the region, I watched thousands of people try to build the connections that would grow their businesses and advance their careers. Most of them worked harder than needed. They certainly didn't lack the talent or drive to succeed in networking, but they didn't have a clear system for making it a routine, productive way to grow their enterprise. Again and again, I saw the same challenge. True professionals know networking is important. They show up, attend events, and introduce themselves to everyone they can. Yet the results are often disappointing: superficial conversations, no follow-ups, and no new promising connections.

Networking feels uncomfortable for many people, sometimes even random. They hope the right conversation will happen at the right moment. Sometimes it does, but why stake your business success on the arbitrariness of chance encounters? But over the years, I also saw people who didn't struggle with networking. They understood how relationships grow and approached networking with intention. They prepared before conversations, listened carefully, and looked for ways to help others. Most importantly, they followed up thoughtfully.

Crucially, their networks grew naturally over time, and their businesses along with them. Because of my experience speaking and writing about networking, the question about artificial intelligence comes up all the time. *How can AI help me network?* The answer is both simple and important. No single prompt will make someone a great networker. No technology can replace curiosity, empathy, and the human ability to build trust. Relationships are built by *people*. But I can't deny that AI is a powerful tool. It can help you prepare for conversations and quickly learn about industries and roles. It can help you reflect on conversations and organize your follow-up. Used thoughtfully, it removes many of the small obstacles that make networking harder than it needs to be. What AI cannot do is build the relationship for you, which is why this book does not focus on one specific platform or tool. Technology evolves quickly. New systems appear every year. Today, people use tools such as ChatGPT, Claude, Gemini, Grok, Llama, and others. Tomorrow there will be new ones. The exact platform matters less than the way you use it. Most modern AI systems share similar capabilities. They help you think, organize information, and explore ideas. What matters is learning how to incorporate these tools into a process that supports real human interaction. That is why I wrote this book. In it, I introduce a simple system for networking in the age of AI, showing you how to use these tools to prepare for and reflect on networking opportunities while keeping the human conversation at the center. It's a straightforward and three-step process:

1. AI before the conversation.
2. Human during the conversation.
3. AI again after the conversation.

Over the course of my career, I've seen small business owners grow their companies through relationships built one conversation at a time. I've also seen professionals advance their careers because someone remembered them, trusted them, and recommended them. But I've also seen talented people struggle simply because they never learned how to build those connections. Networking is not about collecting business cards or adding contacts to a list—it's about building relationships that develop over time. My goal in writing this book is simple. I want to share a practical approach that professionals and small business owners can begin using immediately. You don't need complicated strategies or special software. You *do* need a clear process, curiosity about other people, and the discipline to follow up. Artificial intelligence can support that process. But the real power of networking will always come from something much older than technology: the simple act of being genuinely interested in another person. On the drive home that evening after giving my networking talk, I kept thinking about Tracy's question. How many other professionals feel the same way? They know relationships and networking are crucial, yet they feel uncertain about how to approach them in a world where technology changes everything. The fundamentals have not changed—people still trust people.

Technology can help you prepare. It can help you reflect. It can help you organize your thinking. But the conversation itself, the moment when two people sit across from each other and begin to understand one another, is still deeply human. To illustrate how networking in the age of AI works, this book follows "Meg," a professional and a new business owner, as she begins to rethink her approach to networking. Like many professionals, she is capable and experienced, but she has never been taught a clear system for building relationships. Her journey will feel familiar because we all start exactly where she does, in one way or another.

A NOTE ON AI AND NETWORKING

This book is not a technical manual for using artificial intelligence. AI tools are changing quickly, and every professional uses them differently. Some individuals use customer relationship management systems, while others use spreadsheets or organize their contacts through simple notes and personal workflows. What matters is not the specific AI tools you choose, but whether you develop an effective way to remember people, organize your conversations, and follow up consistently. The examples in this book show how AI can support different stages of networking, helping you prepare before a conversation, organize your thoughts afterward, and draft thoughtful follow-up messages. Later in the book, you will see one example of a Personal Relationship Intelligence System that works without a traditional CRM. It is simply one practical approach for individuals who want a simple structure. Adapt these ideas to fit the tools and systems that work best for you.

The goal of this book is not to teach software, but to understand the dynamics of relationship-building in the age of AI.

CHAPTER 1
THE INFORMATION PROBLEM

The night before the event, Meg Wilson opened her laptop and began to research. Three months earlier, Meg had launched her organizational health and performance consulting practice, focusing on burnout, leadership transitions, and workplace well-being. For more than a decade, she had worked in the health field and understood the science behind employee wellness. She knew how stress, workload, and leadership pressure affect performance and health. Meg had learned early in her career that most workplace health problems don't start in the gym or cafeteria. They start with leadership pressure, unclear expectations, and teams pushed beyond sustainable limits. What she did not understand yet was how to build the relationships that would bring clients to her new practice. Meg knew that if her consulting practice was going to succeed, her networking efforts needed to build a client list. When she registered for a regional business networking event, she decided to prepare in a modern way: by opening an AI tool and uploading the attendee list. Within seconds, the screen filled with information that would have taken hours to gather by hand, such as company summaries, leadership bios, recent press coverage, and industry trends. The AI tool even suggested possible conversation starters tailored to specific attendees.

This would be her first serious attempt at using AI to prepare. One name caught her attention—Daniel Brooks, chief operating officer of a mid-sized technology company. The tool flagged Brooks's company as growing quickly and noted that companies at that stage often explore employee wellness programs to reduce burnout. It surfaced something else, too: a podcast interview in which Brooks spoke openly about long hours and rising stress across his teams. The AI suggested a possible opening line: *Ask about scaling culture during rapid company growth.* Meg leaned back in her chair. In less than ten minutes, she had insights that would have taken an entire week to gather on her own. For the first time since launching her business, networking felt less like walking into the dark.

The First Networking Event

The next evening, she arrived early at the networking event. The room filled quickly. Small groups gathered near cocktail tables and the coffee station. Conversations rose and fell in waves across the room. Meg spotted Daniel Brooks almost immediately. He stood across the room speaking with two people. She waited for a lull in conversation and mustered the courage to step forward. "Hi," she said. "Are you Daniel Brooks? I listened to your podcast interview about scaling culture during rapid growth. You mentioned the long hours your teams were working." Brooks looked at her for a moment, not in an unfriendly way, but he did look surprised. "Yes," he answered slowly. "That interview was about two years ago." Meg felt the ground shift beneath her. Two years ago? The AI had surfaced the podcast, but it hadn't mentioned that it was already outdated. "We actually addressed a lot of that after that interview," Brooks continued. "Ultimately, we restructured how our teams operate." Meg and Dan Brooks exchanged a few polite comments. Then he excused himself to greet someone across the room. Meg stood there for a moment, replaying the interaction. The tool had given her impressive information. But it had also given her something she had not expected. A false sense of familiarity. She knew facts about Dan Brooks, but she didn't know Dan Brooks, the person. For the next several minutes, Meg lingered at the edge of the room. Her instinct was to pull out her phone and search for the next person on the attendee list, to find another insight, angle, or prepared opening. Preparation seemed safer than another awkward conversation. Luckily, she resisted the urge. Instead, she watched how people interacted. From her vantage point, the most natural conversations in the room did not look like they started with research or rehearsed introductions. They began with *questions*, like these:

- "What brings you here tonight?"
- "What kind of work are you doing right now?"
- "What challenges are you seeing in your company?"

Meg walked toward a small group discussing employee retention and workplace stress. This time, she tried something different. "Hi," she said. "I'm Meg Wilson. I work with companies on burnout, stress management for leaders, and leadership transitions during periods of rapid growth. I'm curious what companies are seeing right now when it comes to employee well-being." The conversation opened up to her immediately. One executive described rising burnout in her department. Another talked about the difficulty of keeping employees energized during long development cycles. A third mentioned that his company had tried two different wellness vendors in the past year, and neither had lasted. Meg listened, asking follow-up questions when appropriate, and sharing insights from years of studying burnout and workplace health. And she did these things, not like she was delivering a pitch but as someone genuinely interested in the discussion. Fifteen minutes later, one of the executives handed her a card. "You should send me more information about what you do," he said. "This might be helpful for our team." Driving home that evening, Meg thought about the two conversations. With Dan, she had arrived armed with information. She knew his title, his company's growth trajectory, and a quote from an old podcast. She felt prepared, but that conversation never found its footing. In the second case, she opened with a genuine question. That question opened a door that all her research and AI tools couldn't. The technology she used had not been useless. It helped her identify companies worth meeting. It gave her context on industries and trends, but it also told her something about networking: preparation and connection are *not* the same.

Changing the Playing Field

Artificial intelligence is changing how professionals prepare for conversations, research opportunities, and reach out to new contacts. Today, you can learn about a person's company, career history, public statements, and professional network within minutes. You can walk into a room knowing something about nearly everyone in it. And yet, is networking getting easier? It's not. In fact, it's getting harder in many ways. When everyone has access to the same information, information stops being an advantage. When everyone can generate polished outreach messages and well-researched introductions, those messages begin to sound the same. Preparation has been transformed by technology, but connection hasn't. The real challenge of modern networking is not learning more about people before you meet them. The challenge is learning *how to turn all that information into genuine relationships.* Meg discovered that lesson in a single evening. The moment she stopped relying on what an algorithm told her to say and started asking real questions, the conversations changed. Artificial intelligence will continue to reshape how we prepare for networking. But relationships still begin the same way they always have, with curiosity, attention, and a willingness to listen. Meg walked into that event believing information would give her an advantage. What she discovered instead is the starting point of this book. Information can open the door. You need to build trust to get invited inside.

CHAPTER 2
THE FOLLOW-UP MYTH

The morning after the event, Meg Wilson sat at her kitchen table with three business cards in front of her. Meg picked up the first one. She remembered the conversation clearly. The executive had been thoughtful, curious, and open about the stress his team was experiencing. They had discussed burnout, change management, employees struggling to maintain healthy routines during long product cycles, and the difficulty of finding programs that not only support employee well-being but also help leaders guide their teams through constant change.

When Meg came to the second card, it took a moment longer to recall the conversation. It had been polite and professional—the kind of exchange that feels productive in the moment but fades quickly afterward. Meg picked up the third card. Staring at the name, she realized something uncomfortable: she couldn't remember this conversation at all. Here was another key insight from last night's event: every networking conversation feels meaningful. The energy of the environment makes it seem as though relationships are forming everywhere. But the next morning revealed the truth, that events facilitate introductions, while relationships are built afterward. Meg opened her laptop. The easiest thing to do would have been to send a quick follow-up to all three contacts. In fact, the AI tool she had used the night before could do that instantly. It could generate a clean, professional message in seconds. With a few prompts, it could even personalize the note using the information it had gathered about each attendee. Meg tried that. Within a few seconds, the screen displayed the following polished email:

> *Hi Mark,*
>
> *It was great meeting you at the event last night. I enjoyed learning about your company and would love to stay in touch . . .*

Professional and efficient, the email seemed perfectly acceptable until Meg reread it. It might have hit the right notes, but it wasn't *real*, not in the way Meg needed it to be to make a genuine connection. The conversation she had with the executive hadn't been about introductions or general networking. It had been about a very specific challenge. His company had tried multiple wellness vendors, and none of them had lasted. The discussion had been honest, even a little frustrated. The bland, AI-generated message didn't capture that. Time to try again. She flipped the business card over and read the name again: Mark Reynolds. Meg wrote a new email:

Hi Mark,

I enjoyed our conversation last night about the challenges your team has faced with employee burnout. Your comment about trying several wellness vendors and not finding the right fit stayed with me. Over the past few years, I've helped a few companies approach this differently by starting with small pilot programs instead of large rollouts. If you're open to it, I'd be happy to share what that looks like and hear more about what your team has been experiencing.

Meg paused and read the message again. Her version of the follow-up email felt right, reflecting the conversation they'd had. But she didn't send it yet. First, she opened the networking project she had created last night. In Mark Reynolds's chat, she stored everything related to the relationship: her notes about the company, information about their industry, and now the message she had just written. Then, she gave her AI the following prompt:

Check the tone of this message. Make sure it sounds professional but still personal. Correct any grammar if needed and keep the voice and meaning the same.

The AI suggested a small grammatical correction and recommended changing a phrase to make the sentence clearer. But it kept Meg's wording, tone, and message structure intact.

The technology did not replace her voice; it *supported* it. Meg sent the email. She repeated the same process with the other two contacts. Each message started from her own words and memories from the conversation. The AI helped her stay organized, store correspondence within the project, and ensure the messages were clear and professional. Two days later, Mark replied. Meg saw the notification on her phone while she was finishing a client consultation. She opened the message.

Meg,

I appreciated your follow-up. The idea of starting with a pilot program is interesting. Would you be available for a brief call next week to discuss it?

Best, Mark

Meg smiled. This was promising. The call happened the following week. Mark invited two members of his leadership team to join in. They spoke about employee stress, retention challenges, and the difficulty of implementing wellness and leadership stress-management programs that employees use. For her part, Meg listened more than she spoke. When the conversation ended, Mark observed, "We've had several vendors approach us," he said. "Most of them came in with full presentations and big proposals. Your idea of starting small feels more realistic." Meg smiled again. Two weeks later, Mark's company agreed to test a pilot program with a small group of employees. This was Meg's first corporate client for her new consulting business. The opportunity started at the event, and the follow-up turned it into a real opportunity.

Networking for Genuine Opportunities

Many professionals believe the hardest part of networking is walking into a room full of strangers. A harder challenge follows: transforming a conversation into a relationship. You must remember what people say and thoughtfully follow up on the initial conversation. The best way to do that is to continue the discussion rather than restarting it with a generic message. Artificial intelligence can make parts of this process easier by helping you organize your contacts, store information about the companies you meet with, review your messages for clarity, and remind you when to follow up. Technology allows professionals to reach hundreds of people quickly. LinkedIn inboxes fill with messages that say, "Great connecting with you." Automated follow-up sequences send reminders to stay in touch. Contact lists grow faster than ever before. However, too many connections never become relationships. You've probably experienced it yourself: that perfectly written message that feels strangely impersonal. If your follow-up note is so generic that it could have been sent to anyone, it won't accomplish your objective of getting that next meeting. Why would someone reply to a connection request from someone who clearly didn't listen to the initial conversation enough to remember your needs? AI's efficiency at supporting networking efforts is real, but so is the distance it can create. Relationships don't grow at the speed of software. They grow according to the amount of human attention you show a prospective connection.

One Conversation Continued

Meg saw this clearly in the weeks after the event. The contact from the second business card never responded to her message. The name on the third business card replied once with a polite thank-you and was never heard from again. Only the conversation with Mark persisted. Meg hadn't used the most advanced tools, but the conversation had an aura of reality and connection from the beginning. Meg had walked into the event believing the most important moment would happen in the room. Next, she would learn that the real work of networking begins after you leave the event.

CHAPTER 3
THE TRUST METHOD

Do you know what information without a system is? Noise. Meg Wilson was starting to understand that. The networking event had taught her more than a year of reading business articles ever had. The first event had left her embarrassed. She had approached Dan Brooks, confident she was prepared. She knew his title, the company he worked for, and even referred to a pain point he'd mentioned during a podcast interview. Her face flushed scarlet as she remembered the look on Dan's face—his polite smile and his pause after she mentioned the outdated podcast. Their exchange lasted less than a minute. In a way, she could call the event a moderate success because she met Mark, and with her thoughtful follow-up email, was able to book a consulting meeting for a human resources pilot program. But the other two conversations, while pleasant, had led nowhere. Two hours of introductions, netting three business cards, but only one conversation that connected to the work and a follow-up. Of the other two business cards, she could only remember the person who had given her the second one vaguely, and she couldn't remember the person who had given her the third one at all. Considering these experiences, Meg opened a notebook and wrote two questions:

- *How do I find the right people?*
- *Once I meet them, how do I build a real relationship?*

For a long time, she stared at the page, tapping her pen.

What Works Better than Information

Starting a consulting practice was exciting in theory. In practice, it meant nights sitting alone with the uncomfortable possibility that she might not be very good at this yet. Looking through her chats, she reviewed the last two weeks. First, she noticed how much information she had collected: articles, company notes, LinkedIn profiles, and even industry reports. Enough information to power an intelligence agency.

Yet information had not guaranteed her success in networking. The one conversation that led to anything—the one with Mark Reynolds—had worked because she *stopped* trying to deliver a prepared pitch and had instead walked toward a small group near the back of the room and asked a simple question about their discussion. She had listened, following the conversation rather than steering it. And then she followed up on that initial contact with a personalized email that showed she not only remembered the conversation they'd had about retention struggles but also had a thoughtful solution—a limited-scope pilot program that allowed Mark's company to test her ideas on a smaller scale, less likely to disrupt business operations.

From a Pattern to a System

Meg leaned back in her chair. A pattern was starting to appear. Technology had helped when preparing for an event and analyzing it afterward. It had also provided a careful review of the email she had written to Mark, checking the tone and accuracy and making small grammar corrections. During the conversation itself, however, what mattered most was Meg's curiosity and listening skills as the executives discussed their retention challenges. Meg picked up her pen and wrote three lines in the middle of the page:

- AI **before.**
- Human **during.**
- AI **after.**

That was the pattern that had worked. She circled the lines. What she needed was not more information; she needed a *system.*

Target

Then Meg had another thought, so she turned to a blank page. She not only had the timing wrong for how she should use AI, but she also hadn't been very discerning when it came to the types of people she was approaching. She had been going to random events, approaching random people, who had nothing to do with the work she wanted to build, a consultancy based around human resources issues. She wrote a critical word at the top of the new page: **target**. What was targeting? She brainstormed, jotting her ideas down on the page. *Find the right people. Find the rooms where those people gather.* She glanced at the three business cards again. One good conversation. Two wasted ones. If she had approached these wasted conversations the way she had approached the one with Mark Reynolds, would she have three leads for her consulting business instead of just one? Or was it something to do with her confidence? It was neither of those things. It was the room—the networking event she had chosen, the people she had approached.

Research

Meg wrote a second word in her notebook: **research.**

If she had known who she wanted to meet, she could have used AI to understand their world before the conversation began. Not memorizing facts about them, but understanding the challenges they might be dealing with.

Understand

Meg paused for a moment. A third word came to her, something that she couldn't delegate to any AI tool. She wrote it down: **understand.** This was the part that happened in the room itself. It was asking questions, listening carefully, and responding to what the other person was saying. It wasn't delivering a pitch.

Support

Meg tapped the pen against the notebook for a moment, then added the next word: **support.** Even great conversations fade quickly without follow-up. Relationships grow in the messages that progress the conversation, the details you remember, and the problems you can solve. Meg regarded Mark Reynolds' card again. She remembered reading the message she wrote to him before sending it. She had asked herself a question before pressing send: *If I were standing in front of him right now, would these words sound natural?*

Track

Meg wrote the final word in her notebook: **track.** That third business card from the conversation she couldn't remember brought the importance of this concept home. To turn a networking target into a viable lead, she needed to build a system that could track the following:

- How did the conversation go?
- What mattered most to the person?
- When would she reach out again?

Without tracking these things, relationships disappeared into the same pile of forgotten introductions.

What Does It Spell?

Meg leaned back and looked at the page. Five words:

- **T**arget
- **R**esearch
- **U**nderstand
- **S**upport
- **T**rack.

Meg sat up in her chair, truly excited about the concept she'd created: an anagram that represented the key aspects of networking in the age of AI, one that spelled out an even more crucial concept: **TRUST.** The word had appeared on the page by accident. The more she looked at it, the more right it felt. Every step pointed toward the same outcome. It wasn't a transaction or a quick sale. If followed correctly, this system would produce a relationship strong enough to build enduring trust. Meg closed the notebook and gathered the business cards from the table. For the first time since starting her consulting practice, networking felt less like a gamble and more like something she could learn to do well. The next event would not be random. She would start by choosing the best networking event, the right room. When she walked into that room, she would know exactly what she was trying to build. Information is a foot in the door. Only by building trust can you get inside.

Key Take away: The TRUST Method

The TRUST Method provides a clear structure for networking by organizing relationship building into these five repeatable steps:

- **Target:** Choose the right people and the right rooms. Focus your attention on the environments where the conversations that matter are most likely to happen.
- **Research:** Use AI and available information to understand the person and their world before the conversation begins. The goal is not memorizing facts; instead, enter the conversation with context rather than guesswork.
- **Understand:** Listen, ask thoughtful questions, and pay attention to the other person's pain points, the problems they are trying to solve. Real networking happens when you understand the person in front of you rather than trying to impress them.
- **Support:** After the conversation, follow up in a way that shows you were paying attention. Mention something specific you discussed. A short message with real attention is far more powerful than a long message that says nothing.

- **Track:** Relationships develop over time. Keep notes about the people you meet, what matters to them, and when it makes sense to reconnect. Without this step, conversations disappear into the same pile of forgotten business cards.

Applying the TRUST Method

Try this framework the next time you prepare for a networking event, meeting, or introduction.

Target

Decide who you want to meet before entering the room. Identify the people or organizations connected to the work you want to build.

Research

Use AI or online research to learn details about someone you may meet. Identify their role, industry, and the problems they likely need solved.

Understand

Prepare two questions that help you understand the other person's work, challenges, or goals.

Support

After the conversation, send a short message that refers to something specific you discussed.

Track

Write down the person's name, what you learned about them, and when to reconnect to continue the conversation. Repeat this process, and networking becomes more intentional. Relationships begin to develop with *intention* rather than chance.

Practical Example: Preparing for a Networking Event Using AI

Networking events move quickly. You often have only a few minutes to meet someone and start a meaningful conversation. A small amount of preparation can make those minutes far more valuable. AI can help you prepare before the event and organize your thoughts afterward.

Before the Event

Suppose you are attending a networking event where many participants are executives and leaders from engineering and technology companies. You can open your AI tool and craft a prompt such as:

> *I am attending a networking event with many executives from engineering and technology companies. What are three thoughtful questions I could ask to start meaningful conversations about leadership, team growth, or industry challenges?*

The response might suggest questions such as:

- *"What changes have you seen in how engineering teams operate as companies grow?"*

- *"What leadership skills matter most for engineers who move into management roles?"*
- *"What challenges are you currently seeing in hiring or developing engineering talent?"*

Preparing questions like these helps move the conversation beyond simple introductions and toward topics that matter to the people you meet.

During the Event

Technology can't replace conversation. Your job at a networking event is to focus on the person in front of you. Listen carefully. Pay attention to what they share about their work and their challenges. Notice the problems they mention, the goals they are trying to achieve, and the changes they are seeing in their industry. Thoughtful questions and careful listening create meaningful interactions, even in a brief exchange.

Connecting on LinkedIn

Instead of exchanging business cards, many professionals connect on LinkedIn during or right after the conversation. As the exchange comes to a close, you can simply ask, *"Would you mind if we connect on LinkedIn?"* Most people will say yes. Open LinkedIn on your phone and send the request on the spot. Connecting on the spot has several advantages. You avoid losing a business card. The person recognizes your name and profile immediately. You also create a clear path for follow-up. You can add a short personal note with the request:

Great meeting you at the leadership networking event tonight. I enjoyed our conversation about engineering leadership.

After the Event

After the event, AI can help you organize your notes and plan a thoughtful follow-up. Try a prompt such as:

> *Here are notes from several conversations I had at a technology networking event. Please summarize the main topics discussed and suggest a short follow-up message for someone who mentioned challenges with leadership transitions in engineering teams.*

A strong follow-up message is specific to the conversation you had:

> *It was a pleasure meeting you at the event earlier this week. I enjoyed hearing about the challenges engineering leaders face as teams grow. I came across an article on leadership transitions in technical organizations and thought you might find it useful. I would enjoy continuing the conversation.*

AI organizes the information. You build the relationship.

CHAPTER 4
FINDING THE RIGHT ROOMS

If the TRUST Method was going to work, Meg realized she needed to apply the first step meaningfully. Target. Meg opened her laptop and typed her first question into her AI platform.

List communities, conferences, and professional groups where HR leaders and operations executives from growing technology companies meet.

The response appeared almost immediately.

Startup leadership meetups. Technology operations forums. People operations associations. Scaling-startup conferences. Executive leadership communities.

The list kept expanding. Meg scanned the results slowly. The list was impressive, but overwhelming. AI had answered her question, but it hadn't solved the problem. There were too many rooms. Meg refined the prompt.

Which of these attracts leaders from companies with 50 to 300 employees?

The results narrowed. She decided to get even more specific.

Which of these groups has events in this city in the next three months?

Now the answer was manageable: five events. Meg saved the results in the networking project she had opened for this effort. She found it worked best to organize her work this way: each networking effort with its own project. Conferences, relationships, and potential clients each lived inside separate projects, where information could accumulate over time. Her system allowed the AI to cross-reference everything inside this project. Events, organizations, and people: future questions would draw on all of them. As Meg reviewed the list of events, one speaker's name caught her attention: Daniel Brooks. He was the executive she had approached at one of her earlier events, confidently quoting a podcast interview she thought was recent. It turned out the interview was about two years old. Meg still remembered his awkward pause at her question. Dan had been polite, but the conversation ended quickly. Could she conquer her embarrassment enough to approach him at an event again? She decided she could. And it would be a perfect opportunity to prove to herself she was serious about networking. Meg opened LinkedIn and found Daniel's profile. LinkedIn also showed several mutual connections. Meg copied the names into her networking project; it was more context for the system she was building.

The event, the sponsoring organizations, the speakers, and their networks were now all part of her networking project. One name stood out: Alyssa Chen. Meg recognized her from a leadership conference the previous year. As she reviewed Alyssa's profile, a pattern emerged. Alyssa helped organize leadership events and frequently moderated conference panels. She was also connected to several of the speakers appearing at the same conference where Dan Brooks would be presenting. Meg crafted a new AI prompt.

Based on the speakers at this event, the organizers, and the mutual connections I just added, which person is most likely to have a direct working relationship with Daniel Brooks?

The system reviewed the information inside the project. Alyssa Chen appeared repeatedly in connection with conference organization and speaker coordination. The event page listed her as one of the program coordinators. Meg smiled; this was exactly the kind of cross-reference she could not have spotted easily on her own. Alyssa was not simply a mutual connection—she was part of the *structure* behind the event and a natural bridge to Dan Brooks. Meg wrote a short LinkedIn message.

Hi Alyssa.

I noticed you're connected with Daniel Brooks and saw he's speaking at the Scaling Leadership event next week. I'm planning to attend and would enjoy reconnecting with him if it makes sense. Would you feel comfortable introducing us?

She read the message once. Then again. It was simple, respectful, and easy to decline. Meg pressed send. Two days later, Alyssa replied: *Of course. I'll introduce you at the event.* Meg registered for the conference immediately.

The Second Networking Event

The room was already busy when Meg arrived. Within minutes, she noticed something about the conversations around her. People were discussing the same kinds of challenges Meg had created her consultancy to solve, challenges such as these:

- leadership transitions
- newly minted managers
- communication problems in rapidly growing companies

For the first time at a networking event, Meg did not have to search for relevant conversations because they were already happening all around her. Alyssa caught up with her a few minutes later. "Meg, good to see you again," she said. "Come meet Daniel." Dan turned as they approached. For a brief moment, Meg remembered their earlier conversation. Then she focused on the present moment. "Dan," Alyssa said, "This is Meg. She works with companies on burnout, leader stress management, leadership transitions, and workplace wellbeing during periods of growth." Smiling, Meg added, "I heard you speak recently about preparing first-time managers in fast-growing companies," she said. "What patterns are you seeing most often?" Daniel nodded. This time, he obviously approved of her question. "The biggest challenge is that companies promote strong individual contributors into management roles without preparing them to lead." Meg listened carefully to his words. "That's exactly what I've been seeing," she replied. "Especially when companies double in size within a year." Daniel leaned forward slightly. "Then you've probably seen what happens next." Meg nodded. "Managers who were great performers suddenly struggle because no one taught them how to coach people." Daniel smiled, and Meg felt she had vindicated her previous performance. Later that evening, Meg returned to her project, adding notes from the event about Daniel's comments on leadership transitions and the names of several people she had met for the first time.

Now the system had access to these new data points and could draw on the event, the relationships, and the ideas discussed. Meg paused and thought about what made this event so successful, ultimately realizing she hadn't entered the room, hoping the right conversation *might* appear. Instead, she had entered the *right* room prepared to listen. The difference was small. Yet, it altered the trajectory of the entire evening before the first question was even asked.

Key Take away: Finding the Right Rooms

The first step of the TRUST Method is **Target.** Targeting means choosing where you spend your attention. Many people try to meet everyone everywhere, which creates *activity* but little *progress.* Progress begins when you place yourself in environments where the right conversations already happen. Here's a four-step method to make that happen.

1. Find the Right Rooms

Start with a broad search. Open your AI research tool and ask:

List communities, conferences, and professional groups where [your target audience] gathers.

The first answer will produce a long list. Most entries will not matter. Narrow the field by refining your query, as in:

Which of these attract organizations with [specific industry, size, or stage]?

Then narrow again:

Which of these hold events in my city within the next three months?

Notice that each question moves from general awareness to a short list of places where meaningful conversations become more likely. The goal is not more rooms. The goal is the *right* room.

2. Find the Bridge

After identifying a room, look for people who already belong there. Search LinkedIn for speakers, organizers, and active members connected to that community. Look for overlap between their network and yours. Add those names to your AI research project. Ask a cross-reference question such as:

Based on these speakers, organizers, and shared connections, who is most likely to have a direct working relationship with [the person I want to meet]?

AI can't see your network directly. It works with the information you provide. When you combine your network knowledge with its ability to analyze relationships, new paths often appear.

3. Write the Message

Once you identify a bridge, request a short introduction, keeping your query brief. Explain why you want the conversation and show respect for the other person's time. One thoughtful introduction message will create more opportunities than dozens of cold requests.

4. Choose Rooms Carefully

Networking improves when you stop chasing volume and start choosing environments with intention. The right room changes the conversation before it even begins.

CHAPTER 5
PREPARE WITH INTELLIGENCE

The next day, Meg was halfway through her morning coffee when her phone buzzed. A message from Daniel Brooks appeared on the screen.

It was good meeting you at the conference. I enjoyed the conversation.

Excited, Meg read the message twice. After an awkward first contact, she had redeemed herself and found a new lead, all because she had walked into the right room at the right networking event, prepared to listen and understand what mattered to the man she wanted as a client. Meg opened her laptop and navigated to her networking project. Inside the project, she already had several pieces of information saved about last night's event: organizers, speakers, mutual connections, and notes from her conversations. Meg added new information to the chat named "Daniel Brooks." She added his message, notes from the conversation, and two recent articles about his company's growth.

Asking the Right Questions

Now her system had *context*. This was the advantage of organizing her networking efforts as a project. When Meg asked a question, the system would not have to search the entire internet for relevant information. It would analyze the information already in the project and build answers from those sources that were both relevant and validated. Meg typed her first prompt.

What are the three operational challenges a COO of a rapidly growing technology company is likely dealing with right now?

A response appeared.

Hiring and onboarding leaders fast enough to support growth. Maintaining culture during rapid expansion. Preventing burnout across teams.

Daniel had mentioned several of those issues during their conversation. Meg typed some additional notes into Daniel's chat.

Follow up with Daniel in two weeks. Send the article about leadership transitions. Send me a reminder email in two weeks.

Meg typed another prompt.

What questions could someone ask Daniel Brooks that would help him talk about these challenges?

Her AI generated several possibilities:

What systems help first-time managers succeed when teams grow quickly? How do you keep communication clear as teams double in size? What leadership habits become most important during rapid growth?

Meg did not copy the questions—not yet. Instead, she studied them. All the questions focused on Daniel's problems rather than on pitching her services. That was the shift she had been learning to make. The goal of networking wasn't to impress people, but to *understand* them. Meg started a new chat inside the project. She wrote the four lines:

- *understand the person*
- *understand their environment*
- *understand the problems they need solved*
- *arrive curious*

That was it: her preparation system, which she would undertake before networking events, meetings, and targeted conversations. Meg decided to test the system again, typing another prompt:

What decisions does a COO in Daniel Brooks's position make regularly?

The response appeared:

Hiring leadership roles. Building operational systems. Maintaining company culture during expansion.

Meg typed a short note beside the response: *operational leadership.* Her consulting practice focused on helping organizations build healthier performance systems by reducing burnout, supporting leadership transitions, and keeping teams sustainable during periods of rapid growth. If Daniel's company was growing quickly, burnout was a real possibility.

Meg typed another prompt.

What signals indicate that a company at this stage is beginning to experience employee burnout?

The answer:

Rising turnover. Managers struggling to support their teams. Communication breakdown between departments.

Meg paused. She remembered something Dan had said at the conference: "We've doubled in size this year. Communication is getting harder." Meg added the quote to Daniel's chat project. Now the system connected Daniel's comment, the company articles, and the operational patterns of growing companies. Preparation was no longer abstract—it was becoming specific. Meg tried another prompt.

What leadership topics might Daniel Brooks be thinking about right now?

The AI's response:

Leadership development. Manager training. Scaling communication systems.

Meg frowned. The answer made sense, but it was too general for her purposes. She refined the prompt:

Based on Daniel Brooks's company growth, the articles about their expansion, and the operational responsibilities of a COO, what leadership challenges are most urgent for him right now?

This time, the AI responded differently:

Preparing first-time managers. Preventing burnout among senior engineers. Building clearer communication between teams.

Okay, she was getting somewhere. The difference between this and the previous response was *specificity*. The first prompt she'd tried asked what Daniel *might* be thinking about, but she didn't provide any parameters. On her second attempt, she added specific, relevant things for the AI to reference in its response. With the second prompt, the AI delivered a more relevant and targeted answer.

The Third Networking Event

A week later, Meg attended yet another leadership event, her third since she had been trying to systemize AI for her networking purposes. In the past, she would have spent the first fifteen minutes thinking about how to introduce herself. Now she walked into the room looking for something else—problems, such as these:

- leadership problems
- growth problems
- communication problems

Near the event's coffee bar, a small group was talking. "We promoted three engineers into management roles," said Rajiv Gupta, a chief technology officer of a mid-size computer chip design company. "They're excellent technically, but none of them has managed people before." Meg immediately recognized the challenge: first-time managers. It was the same challenge her AI system had identified earlier. She joined the group. At a pause in the conversation, she asked, "What kind of support are you giving those three newly promoted managers right now?" Rajiv looked thoughtful. "Honestly, not much yet," he replied. "We're still figuring that out." Meg almost fell back into her old habit of describing in detail the work she did with leadership teams. Instead, she asked, "What has been the hardest part for the first-time managers so far?" Rajiv thought about it. "Giving feedback," he finally answered. "They're great engineers, but managing people is new territory." He went on to explain further his challenges with first-time managers. Meg listened more than she spoke. When the group eventually separated, Rajiv handed her his business card. "I'd like to continue this conversation," he said. Later that evening, Meg returned to the networking project. She added a new chat to her networking project with Rajiv Gupta's name, the company he worked for, and notes from the conversation. The system connected the leadership problem, the event, and the growing company. Each new chat she added to her project made the next conversation easier to prepare for. Meg looked at the screen for a moment. Preparation had changed the way she entered conversations, making her infinitely more efficient and successful. No longer trying to impress people, she was trying to understand them instead. The room hadn't changed. But Meg had.

Key Take away: Preparing for the Conversation

Preparation changes how a conversation begins. When Meg prepared for her second meeting with Daniel Brooks, she wasn't trying to predict the conversation, like she had the first time around. She was trying to understand the context. And in her recent conversation with Rajiv Gupta about first-time managers, she had anticipated that might be a topic he was interested in based on the analysis she conducted with her AI system before the event. Preparation enabled her to make the leap from guessing to understanding.

The Preparation Rule

After her most recent networking conversation with Rajiv, Meg wrote four lines in her notebook:

- understand the person
- understand their environment
- understand the problems they need solved
- arrive curious

These four concepts guided how she prepared for every conversation that followed. They helped her move beyond titles and biographies and begin thinking about the world the other person worked in everyday, to understand not just their challenges but their *context.*

Use AI to Build Context

AI tools can help you quickly build context. Start by gathering a few pieces of information in advance. Good sources include LinkedIn profiles, company websites, articles, interviews, and posts where they discuss their work. Then begin with a broad research prompt, like this:

What challenges do engineering managers typically face when they move from individual contributor to leadership roles?

This gives you a general landscape. Then you can refine the question using the person's specific situation:

What challenges might Rajiv's team face as engineers transition into leadership roles inside a fast-growing engineering organization?

Notice that the second question moves from a general role to a real situation. Both types of questions are useful. Broad questions provide orientation, while specific questions create understanding.

Prepare Questions, Not Speeches

Preparation is not about writing the perfect canned introduction. It is about identifying two or three thoughtful questions to help you better understand the other person's work. Questions about decisions, pressures, and challenges usually lead to richer conversations than questions about titles or responsibilities. Here are a couple of examples:

- *What surprised you most when your engineers first started moving into leadership roles?*
- *What problems appear once someone becomes responsible for a team instead of their own work?*

These questions invite explanation rather than short answers, enabling you to gather more information while showing that you are listening to your lead's challenges in a thoughtful and curious way. The purpose of research is simple: to help you arrive informed enough to ask thoughtful questions and open a meaningful conversation. Research should *never* replace curiosity.

CHAPTER 6
ENTER THE CONVERSATION

Once again, Meg sat at her desk with her networking project open. The project had been growing for weeks. It contained notes from conversations, her notes on leadership patterns, and articles about companies expanding quickly. Every event she attended added more information to the system. As she reviewed the notes, she saw a pattern—most networking conversations began the same way, something along the lines of "What do you do?" Meg had asked this question herself hundreds of times. The answer yielded a job title, a company name, and maybe a specific department. Sure, that was all useful information, but it wasn't a *real* conversation, the kind that bonded people and turned casual meetings into viable leads. Meg began to understand the problem with leading with "What do you do?" The question focused on identity, which pointed backward to a resume. Activity questions pointed *forward* toward active problems and challenges. And as she already knew, listening to people's challenges yielded better results. She typed a prompt into her networking project:

Based on the leadership challenges we've been tracking in this project, what opening questions might help a senior manager talk about what they are dealing with right now?

The response focused on work in progress rather than job titles:

Questions about projects. Questions about challenges. Questions about what was currently taking someone's attention.

Meg further refined the prompt:

If someone at a leadership event is managing a fast-growing team, what opening question might help them describe the most important challenge they are facing right now?

The answer was simple:

What part of your team's growth is taking most of your attention these days?

Meg copied the question into her notes and wrote two variations beneath it:

- *What problem has been the hardest to solve recently?*
- *What part of your work has been the most interesting lately?*

All three questions moved the conversation in the right direction: toward *activity*—the problems and challenges a lead was facing.

The Three-Step Opening

Meg opened a new note and wrote a short framework at the top of the page. She called it the **Three-Step Opening,** and it looked like this:

1. Ask what the target was working on.
2. Listen carefully for the problem hidden inside the target's answer.
3. Ask a follow-up question about that problem.

Three-Step Opening Example

To make the framework easier to remember, Meg wrote a short example beneath it.

Step 1

Meg: What kind of projects are taking most of your attention right now?

Networking Lead: We're expanding into two new markets this year.

Step 2

Step 2 was the moment most people missed. But Meg wouldn't. She imagined herself listening to the problem underlying her networking lead's response: expansion. While expanding was a great problem for a company to have, it often entailed hiring pressure, leadership gaps, and communication challenges.

Step 3

Step 3 would allow Meg to build on that observation.

Meg: What part of that expansion has been the hardest so far?

The conversation would move forward from there.

Three-Step Opening in Action

But would it work in real life? Meg was ready to try it at an event that evening. When she arrived at the venue, groups of people were already talking in small circles across the room. These gatherings were fluid; conversations started, dwindled, and people drifted between groups. While hovering at the back of the room, she heard a snatch of conversation that particularly interested her. "We're opening two new locations next quarter," a man said. "Leadership capacity is becoming a real issue." Meg had been anticipating exactly this type of challenge, thanks to consulting her AI system in the hours leading up to the event. Growth. Leadership gaps. These same themes had appeared repeatedly inside her networking project. She edged closer to the group, and during a conversational lull, she said to the man, exactly as she had practiced, "That sounds like an exciting moment for your company. What part of the expansion is taking most of your attention right now?" Mike Delgado, an HR director for a semiconductor manufacturing company, turned toward her. "Hiring the right managers," he said. "We need people who can lead teams, not just run projects." Meg nodded. "What has been the hardest part of that so far?" Mike hesitated. Meg wondered if her question had been too direct. But he answered, "Finding people who know how to coach their teams. Our technical talent is strong, but leadership skills are harder to find." Conversations don't always unfold directly. Sometimes the other person needs a moment to think. Meg and Mike discussed leadership recruitment for several minutes. He described the pressure of expanding quickly while trying to maintain culture across multiple teams. Meg listened carefully, asking probing questions to better understand his problem. When she eventually mentioned that her consulting firm provided leadership development advice, her services that could be of use to Mike emerged organically. As the conversation wound down to a close, Mike handed her his card. "We should talk again," he said. Another win for Meg and her AI networking system. It felt like the best, most natural conversation she'd had with a promising lead yet.

Three-Step Opening Results

After the event, Meg returned home to her laptop and wrote notes about the conversation into a new chat about the experience. These notes included what Mike had said about the company's expansion, his leadership hiring challenges, and the leadership gap he had described. The AI system immediately linked those observations with the patterns she had been tracking across several events. Meg closed her eyes and further analyzed the results. As far as the event itself and the type of people who attended, there was nothing very different about the evening from the previous events. The difference had been the *question* she chose to ask. That, she decided, was what made this event her best effort yet.

Key Take away: Understanding the Problem

In the first five chapters, we looked at ways to open conversations and find the right people at the right events. Chapter 6 shows what to do next. Activity questions help people move beyond titles and begin getting down to business.

Once the conversation opens, your goal is simple: understand the challenge often hidden in your lead's answers.

The Three-Step Opening Reviewed

Meg learned to use this simple pattern whenever she met someone new:

1. Ask what the person is working on.
2. Listen for the problem behind the answer.
3. Ask a follow-up question about that problem.

The first question opens the conversation. The second step requires attention. The third step deepens the conversation. This pattern sounds simple, but most people skip the second step. They move directly from an opening question to their own story, opinion, or advice. Understanding begins when you slow down long enough to hear the problem inside the answer.

Listen for the Signal

When people talk about their work, they often reveal difficulties indirectly. The signals appear in small phrases, like these:

- *That part is harder than it sounds.*
- *We're still trying to figure that out.*
- *That's where things become complicated.*

When you hear a signal like this, resist the instinct to move on. Pause and ask a follow-up question, such as the following:

- *What makes that difficult?*
- *How does that usually show up in your team?*
- *What happens when that situation appears?*

These questions invite people to describe the pressures behind their work rather than the structure of their job.

From Explanation to Experience

Your understanding of your leads' challenges will improve when you advance the conversation from generalities to their specific situation. Instead of asking for opinions about leadership, ask about situations. Instead of asking what someone believes, ask what happened the last time a challenge appeared. For example, you can ask questions like:

- *What happened the last time someone on your team struggled in a new leadership role?"*
- *How did your team respond when that problem first appeared?"*

Specific situations reveal patterns that broad descriptions often hide. Information is power. The more information you gather about your leads' challenges, the better position you will be in to help them.

Stay with the Problem

When networking, many fall victim to the trap of jumping too quickly to offering advice or solutions based on their own experiences. These kinds of simplistic analyses aren't likely to build rapport and trust with your lead. Only when you delve beneath the surface will you gather—and be able to provide—valuable insights. That beneath-the-surface information often looks like this:

- Your lead pauses and adds one more detail.
- Your lead clarifies what makes the situation difficult.
- Your lead describes what they have already tried.

Understanding grows when you allow the other person to surface more information about their challenges before you try to respond. That's the moment when the conversation becomes genuinely interesting.

CHAPTER 7
WHEN THE PLAN DOES NOT WORK

Meg opened her networking project two hours before the next event. After four events, her system had grown into something more useful than notes, including patterns, inferences, and analyses extrapolated from dozens of conversations. Leadership struggles inside expanding teams. Hiring pressures. Technical experts advancing into management roles for the first time. A chime alerted Meg to a new email. Opening it, she saw that Dan Brooks was following up on an article she had sent. Her network was no longer a list of people. It was a series of conversations continuing over time. Meg had started thinking of her networking project as more than just a system—it was her preparation partner. She typed a new prompt into her running project.

Based on the last twenty conversations recorded here, what leadership challenges appear most often in growing companies?

The system grouped the answers into several themes:

- *rapid team expansion*
- *new managers without prior leadership training*
- *technical specialists promoted to leadership roles*

Meg studied these challenges for a moment. Then she typed another prompt:

If someone at a leadership event is dealing with growth, what opening question might help them describe the real challenge they are facing right now?

Her AI generated several new suggestions:

- *What part of your team's growth has been hardest to manage recently?*
- *What leadership challenge has surprised you the most this year?*
- *What problem is taking more time than you expected?*

Meg copied the questions into her notebook. Her preparation was complete; her questions felt solid. If someone mentioned hiring or leadership development, she had ideas about how to continue the conversation.

The Fifth Event: The Bay Area Leadership Conference

As usual, the room filled quickly with people, taking advantage of networking opportunities. People gathered around tall reception tables. Small groups formed near the entrance, proliferating through the space all the way to the back. Meg sauntered through the room, looking for a group to join. As she walked, she listened first. At previous events, people had talked openly about problems related to rapidly expanding teams, new managers, and communication difficulties as companies scaled. But the statements at this event were different:

- *I just moved into strategy.*
- *I joined the executive team this quarter.*
- *I'm running West Coast operations now.*

Meg waited for someone to mention a project or challenge. Instead, she heard job titles, promotions, and other career updates. In other words, these people were reciting their resumes aloud to each other rather than discussing their challenges. Meg finally joined a small group near the center of the room. "What projects are keeping your team busy right now?" she asked when there was space in the conversation. One man answered immediately: "A little bit of everything." Another nodded. "Same here." The conversation drifted back toward introductions. Meg stepped away and approached another group. "What part of your work has been the most interesting recently?" she asked. A woman thought for a moment. "Hard to say. Things move quickly." The answer shut off the conversation before it really began. Meg walked toward the back of the room and stopped near a window. As usual, her preparation had been thoughtful, and she was sure that her questions had been strong. But the room wasn't responding the way she predicted it would. She made up her mind to try again with another group. Then she stopped herself—repeating the same question to ten different people would not change the tone of the room. It probably wouldn't.

She stood quietly and watched the conversations around her. People were speaking comfortably in their groups. Laughter rose from one table near the bar. Across the room, someone introduced two colleagues to each other. It was a lively event. But the discussions stayed surface-level—titles, promotions, and career milestones—no problems or challenges she could leverage to develop a new lead. Meg began to doubt herself. Had she prepared incorrectly? Had the questions she practiced been the wrong ones for this room? She hovered by the window for another minute—listening—unsure whether she should try again or simply observe the evening from the edges. A few moments later, Meg noticed someone standing several feet away. The woman looked in Meg's direction, then glanced back toward the group Meg had spoken with earlier. After a short pause, the woman walked over. She wore a conference badge and held a small notebook. "You asked an unusual question earlier," the woman began. Meg perked up, intrigued to hear what she had to say. "You asked that group what projects they were working on," the woman said. "Most people here ask what company someone works for." Meg smiled. "I've been experimenting with different ways to start conversations." The woman nodded. "I noticed." They introduced themselves. The woman's name was Priya. Priya told her story. She worked at a product design company that had a growth spurt over the past few years. For most of her career, she had been a product strategist, but three months earlier, she had been promoted to lead a large team. "I spent ten years solving product design problems," Priya said. "Now I spend most of my time helping other people solve them." "What has been the hardest part of that change?" Meg asked, after listening carefully to Priya's statements. Priya answered immediately. "Learning when not to solve the problem myself," she said. "My instinct is to jump in and fix things. But if I do that, my team never develops." The conversation continued, progressing beautifully. Priya described the tension between expertise and leadership. Meg asked a few questions but mostly listened. When Meg eventually mentioned her work with leadership development, the connection felt natural. Organic. Priya had already described the challenge herself. Before leaving the event, Priya handed her card to Meg. "We should continue this conversation," she said.

A New Way of Using the System

Later that night, Meg returned to her networking project.

Instead of immediately recording the conversation, as she usually did, she typed a different prompt:

Tonight, the questions we prepared did not work for most people at the event. Why might that happen?

The system responded, citing several possibilities.

- *The audience may be focused on introductions rather than problems.*
- *The event environment may encourage status updates rather than reflection.*
- *The first stage of conversation may require more observation before entering discussion.*

Meg read the responses, then typed another prompt:

If a room is focused on introductions, how should my approach change?

The system suggested two adjustments.

- *Listen longer before entering conversations.*
- *Expect fewer conversations, but deeper ones.*

Wow. That was certainly food for thought. Next, Meg added her conversation with Priya to a new chat in the networking project. She named it Priya. A few minutes later, she closed her laptop to think. The room had been full of people introducing themselves. One person had decided to walk across it and start a conversation with me.

Key Take away: When the Plan Doesn't Work

Preparation improves your chances of having a meaningful conversation. It does *not* guarantee one. Sometimes you arrive in a room that feels different from what you expected. People exchange quick introductions. Conversations stay brief. Questions receive short answers. In these situations, your goal shouldn't be to force the conversation. Your goal is to read the room.

Reading the Room

When Meg entered the event in Chapter 7, she quickly realized that the environment was not responding as she expected.

People were introducing themselves, but they weren't asking each other about their challenges. These were surface-level exchanges. This isn't unusual in professional settings. But great networkers can and do adapt when people behave unexpectedly. Instead of pushing harder, observe longer, keenly noticing how people interact, which conversations last longer than others, and which questions make people pause and think. Adapting to the room is more effective than trying to control it.

Stand Out by Asking Real Questions

In many professional events, people read from the same script: name, role, and company—almost as if these were the only things about them that mattered. The person who asks thoughtful questions stands out, not because they are trying to be impressive, but because curiosity changes the tone of the conversation. Even though Meg asked thoughtful opening questions, people didn't answer with information about their challenges. Discussions stayed brief and identity-oriented. It didn't seem like she was getting anywhere, which Meg found frustrating. But that wasn't the end of the story. Priya approached her later after overhearing Meg's questions, which were different from the staid introductions happening around the room. Sometimes, the most important conversation at an event begins after the room quiets down.

Reflect After the Event

After a difficult or confusing event, reflection matters. Instead of assuming the plan failed, review what happened. Ask yourself questions, such as:

- *What did the room feel like?*
- *Which conversations lasted longer?*
- *Which questions worked?*
- *Which ones did not?*

AI tools can also help with this reflection. Describe the situation and write prompts along the lines of these:

- *What might explain why the room responded this way?*
- *What adjustments could improve the approach in a similar event?*

Reflection turns an uncomfortable experience into a learning moment.

Stay Patient

Networking events are dynamic; not every room will respond the same way or in the way that you expect. Not every conversation will open up immediately or reveal common ground on which you can build a solid relationship. With observation, careful questioning, and reflection afterward, your networking efforts will improve over time. And sometimes, it's that conversation that happens when you least expect it—when the event is almost over or when approached by someone who overheard you speaking with someone else—that will matter the most.

CHAPTER 8
WHEN THE CONVERSATION COMES BACK

Meg was reviewing notes from the leadership conference when she heard a chime indicating she had a new LinkedIn message. It was from Daniel Brooks—he was starting to feel like an old friend. She read the message—then read it again.

> *Meg, I've been enjoying our conversations. The question you asked about first-time managers stuck with me. Would you have time for coffee sometime next week? I'd like to continue our chat.*

Other conversations at networking events had gone well, and through following up, many had progressed to leads. But this was different. Dan Brooks had reached out to *her*, clearly interested in her consulting services. After responding to his message with some available times to meet, Meg opened her networking project, determined to make this next interaction the most successful to date. She navigated to the chat marked with Dan's name. Every new person she met now had their own chat in the networking project. Every meeting, note, and conversation lived inside it. Inside Dan's chat were the notes Meg had written after every conversation, both in-person and online. It also had her notes on Dan's background and the articles they had exchanged.

Daniel Brooks, COO, Horizon Systems, Leadership conference introduction through Alyssa Chen Conversation highlights: Scaling teams quickly after product growth. First-time managers are struggling with communication. Having difficulty transitioning strong individual contributors into leadership roles.

Meg studied her notes. At the conference, the conversation had moved quickly. Daniel had mentioned the challenge in passing while describing the company's recent growth. They had exchanged messages about expansion-related challenges and their impact on human resources. But now the conversation was turning toward Dan's specific challenges and how Meg could help with them. And returning conversations required different preparation than first conversations. Meg copied one line from her notes and entered it into the chat.

Daniel mentioned: 'Our biggest challenge right now is that great engineers become managers overnight and no one teaches them how to lead.

Based on this statement, what leadership challenges typically appear when technical employees transition into management roles in fast-growing technology companies?

The system generated a list of what it said were common challenges:

- *delegating work effectively*
- *lack of training in communication and feedback*
- *managing former peers who are now direct reports*
- *balancing technical work with leadership responsibilities*
- *building trust and authority without formal leadership experience*

Meg read through the list. Once again, her "preparation partner" had come through for her because the AI's answers made sense. It wasn't telling her Dan's exact problem, but *showing* her the landscape around the problem he had already described. She typed another question:

Based on Daniel's comment, what questions could help someone explain their leadership challenge more clearly in a conversation?

The AI generated a new list of questions:

- *When engineers become managers at your company, what part of the transition seems hardest for them?*
- *Do new managers struggle more with people issues or with shifting their own work priorities?*
- *Have you experimented with any training or mentorship programs yet?*

Meg reviewed the questions. They weren't resume questions; they were *activity* questions—exactly what she'd been looking for. They were questions that pointed to problems, inviting exposition rather than dead-end answers. Underneath them, she wrote herself a reminder: *Listen for specific examples.* Meg closed her laptop, feeling a surge of pride. Someone reaching out after an event meant something: her efforts had really landed. Networking was beginning to feel less like a series of introductions and more like a series of returning conversations.

At the Coffee Shop

The following week, Meg arrived at the coffee shop where they had agreed to meet. Daniel Brooks walked in a few minutes later, spotting her and waving.

"Meg," he said, shaking her hand. "Good to see you again." They ordered coffee and sat down. At first, they reminisced about the conference itself—the sessions, speakers, and the long lines for coffee between panels. Then Daniel returned to the subject that had started their previous conversation. "We're growing quickly," he said. "But growth creates strange problems." Meg listened. "Our best engineers are becoming managers overnight," he continued. "They're brilliant technically, but suddenly they're responsible for teams." Meg nodded. "What part of that transition seems hardest for them?" she asked. Daniel paused. "Delegation," he said. "They're used to solving everything themselves. When they become managers, they still want to jump in and fix every problem personally." Meg asked, "Do they struggle more with the people side or with shifting their own work priorities?" "Both," Dan replied. "But the people side surprises them the most. Giving feedback. Handling disagreements. That kind of thing." For the next half hour, Daniel described situations his company had encountered over the past year, including a new manager struggling to supervise former teammates, a project lead reluctant to delegate critical work, and a talented engineer who preferred to solve technical problems rather than guide a team. These were the types of details that never appeared in routine conference conversations. At events, people spoke about *ideas.* In meetings, they spoke about *problems.* Meg listened carefully. Occasionally, she asked another question to keep the conversation moving forward. Finally, Dan glanced at his watch. "I should head back," he said. "But I'm glad we continued this." They lingered outside the cafe for a moment before parting. "At some point, we should talk about bringing you in to work with our leadership team," Daniel said. "That would be great," Meg replied. The conversation had started as networking. Now it was beginning to look more and more like a client relationship.

Following Up

After coffee, Meg typed notes about her meeting with Daniel Brooks into his chat within her networking project, recording the new information she had gathered.

Leadership transition challenges. Delegation issues among first-time managers. Need for structured management training.

Below the notes, she added a reminder to follow up in three weeks and input a request in the chat to send her calendar and email reminders. Across the screen, the Daniel Brooks chat now contained multiple conversations. From crowded networking events to a quiet table in a coffee shop, Meg was getting closer to transforming Dan Brooks from a lead to a client. She was confident that their conversation would further progress at the next meeting.

Key Take away: Preparing for the First Sit-Down

A first sit-down meeting is different from casual conversations at events and text messages. In a first meeting, you try to understand the person and the environment around their work, beginning with preparation and curiosity. In a sit-down meeting, you already have context. You know what the person said during initial contacts—their challenges, and the questions that remained open. Preparation now starts from that point.

Start with Their Words

Meg prepared for her next meeting with Daniel Brooks differently from how she had for their first conversation. Instead of re-researching his role, she opened the chat and returned to the moment when he described the challenge his engineers faced as they moved into leadership roles, paraphrasing his words into her research system. Then she asked the AI a new question.

What challenges might appear when engineers inside Daniel's team move from individual contributor roles into leadership?

Using the person's own words offers a more effective starting point than a generic description of their role. It keeps the preparation anchored in the real conversation that already took place.

Refine the Preparation

After starting with the person's own words, refine the question further. Meg did that through these questions:

- *What situations might Daniel's team face when a strong engineer becomes responsible for managing other engineers?*
- *How might teams respond when someone moves into leadership for the first time?*

Each step brings the preparation closer to the real situations the person might face. The goal is not to predict the conversation, but to arrive ready to explore the problem more deeply.

Events and Meetings are Different

Meg noticed an important difference as her conversations developed. At events, people spoke about ideas. In meetings, they spoke about *problems.* Event conversations tend to be short and exploratory, whereas meeting dialogues are more detailed. People describe real situations, decisions, and challenges they are facing.

Preparation for meetings should reflect this shift. Instead of preparing broad questions about ideas, prepare questions about *situations.*

Continue the Thread

A strong meeting conversation often begins by referring back to the initial contacts, including the following:

- restating the problem
- asking what has recently changed
- exploring how the situation continues to develop

Here's an example of how Meg could do that with Dan: *"When we spoke at the event, you mentioned your engineers were beginning to move into leadership roles. How has that transition been going?"* Returning to the thread shows that you listened carefully and remembered what mattered. When someone sees that you remember the problem they described earlier, the conversation usually moves deeper very quickly.

CHAPTER 9
THE CONVERSATION YOU DO NOT HAVE TO WRITE DOWN

Two weeks after the coffee meeting with Daniel, Meg got an email from someone whose name she recognized right away: Priya Natarajan. They had met briefly at the Bay Area leadership conference. Priya was the person who approached Meg after noticing the unusual question Meg had asked the group. Meg remembered the moment clearly. Priya had listened for several minutes before stepping forward and saying, "That question you asked earlier. . .I've been thinking about it all afternoon." At the event, they had spoken for over twenty minutes. Now Priya was expanding their contact, beginning her message by referring to their first conversation. Priya's message continued: her company was growing quickly. Engineers were stepping into management roles faster than anyone expected. She wrote that she thought Meg might have a useful perspective on those issues. She ended the email with a simple request: *Would you have time for a virtual call next week to exchange ideas?* Meg scheduled a Zoom meeting for the following Tuesday. Then she opened the AI platform where she kept her networking projects. Priya already had her own chat in the networking project that Meg had created when they first met. In that chat were the notes Meg had recorded from that initial contact. She located and reread them.

Priya Natarajan, Director of Product. Previously a product strategist. Conversation highlights: Rapid product expansion, engineers transitioning into management roles, and communication challenges between product and engineering teams.

Meg began preparing for their first formal meeting, scheduled via Zoom. First, she searched for public information about Priya's work, finding two conference talks and a podcast interview from the previous year that appeared online. In the podcast and both conference talks, Priya spoke at length about scaling product teams and managing fast growth. Meg added the transcripts to Priya's chat and asked a question:

What leadership challenges does Priya describe most often in her writing and talks?

In its response, the AI highlighted the subjects that seemed to be of most interest to Priya, based on the transcripts:

Priya repeatedly mentioned tension between product deadlines and engineering workloads. She also described the difficulties new managers faced when promoted directly from technical roles.

Further reading the transcripts, Meg noticed something: in one of her conference talks, Priya described a specific incident in which a new manager was delaying product releases because he was still trying to solve technical issues personally rather than delegating them to his team. Then she asked the AI another question:

> *Based on Priya's speeches and the example she described about new managers delaying releases, what questions could help her explain how this problem appears inside her company?*

The AI system generated several possibilities:

- *When engineers become managers on your teams, how quickly are they expected to stop doing technical work themselves?*
- *Do product deadlines create pressure for managers to step back into coding instead of leading?*
- *Have you seen delays appear because new managers are still solving problems personally?*

Wow, Meg thought, these questions were perfect, as they were specific to Priya's situation. Preparatory questions built on information that was both relevant and recent clearly outpaced those garnered from general research.

Meg's Meeting with Priya

When Meg logged in to set up the Zoom meeting, she enabled the auto-transcription feature to record it. Most meeting platforms have versions of this, and third-party screen recording tools are also available. Meg enabled screen recording for online meetings whenever permitted. She always asked for permission before recording. Most people appreciated the idea once she explained that it helped her focus completely on the discussion, while having something to refer back to later. Priya joined the call. They exchanged greetings and spoke briefly about the leadership conference before returning to the problem that had brought them together. "Our teams are expanding quickly," Priya said. "But our engineers are becoming managers faster than they're ready for." "What happens when they're promoted?" Meg asked. Priya thought for a moment. "They keep coding," she admitted. "They feel responsible for solving every problem themselves." Meg nodded. That made sense. People tended not to change the way they did things until they had to. "Does that ever affect product deadlines?" she asked. "All the time," Priya chuckled. "A manager will delay a release because they're still trying to fix the code themselves instead of guiding the team."

For the next forty minutes, Priya described the situations her teams were encountering:

- managers struggling to step away from technical work
- product teams waiting for decisions that leaders were hesitant to make
- engineers who had been excellent contributors but had never been taught how to guide others

Meg noticed something about the way Priya described the situation—she hesitated whenever she mentioned a struggling manager and leaned forward when discussing communication challenges between teams. These were two situations she obviously found difficult and cared very much about.

Processing the Meeting

After the call ended, the meeting software processed the recording. Within seconds, the transcript appeared. Meg added it to Priya's chat. Then she asked the AI this question:

> *Based on this transcript, what were the three specific leadership challenges Priya described during the call?*

The system analyzed the conversation. It found three trends:

- *new managers coding instead of leading*
- *missed deadlines due to delayed decisions*
- *communication lapses between engineering managers and product leaders*

Meg asked one more question: *Based on these challenges, what follow-up actions would help Priya move this conversation forward?* The AI system responded with these suggestions from all the information to date:

- *Share examples of leadership transition programs used in fast-growing software teams.*
- *Introduce Priya to a company that offers mentorship programs for engineers transitioning into management.*
- *Send Priya an article about helping technical leaders step away from coding responsibilities.*

Those ideas were great. She added them to her main networking project chat.

Changing How She Thought About Conversations

Meg's AI preparation partner had started to change how she thought about conversations. She no longer thought of a series of contacts as isolated moments. They were part of an ongoing conversation, a relationship unfolding. This workflow had become part of her system. She thought of it as the **Conversation Loop.** Before closing the project, Meg added one more note to herself: *Follow up with Priya in two weeks.*

Key Take away: The Conversation Loop

As your conversations become more important, one challenge will make itself clear: you can't remember everything. Your brain has too many other things to do. That's why meeting details get lost. Key moments fade from your memory. Important insights disappear like a vapor. In Chapter 9, we discovered a system for solving this: the Conversation Loop.

The Conversation Loop

The Conversation Loop has three steps.

1. Prepare before the conversation.
2. Focus completely during the conversation.
3. Capture and analyze after the conversation.

Notice how each step supports the next.

1. Preparation helps you ask better questions.
2. Full attention helps you understand the problem clearly.
3. Capture allows you to return to the conversation with accuracy.

Most people focus only on the first step. The real advantage appears in the third.

1. Capture the Conversation

Most virtual meeting platforms include recording, transcription, and AI note-taking tools. If the virtual meeting platform does not include these features, external AI tools can connect to the meeting, record the session, generate a transcript, and automatically produce notes. Use these tools, but always ask for permission. In most cases, people will allow themselves to be recorded for these limited purposes.

May I record this conversation so I can focus and take better notes afterward? Recording removes the need to take detailed notes during the conversation, allowing you to stay fully present. When you aren't focused on the act of taking notes:

- You can listen more carefully.
- You can follow the problem.
- You can ask better questions.

After the meeting, download the transcript and add it to your networking project chat. It is the most accurate record of what happened.

2. Use the Transcript as Input

Once you've placed the transcript in the potential client chat, you can craft focused prompts, like these:

- *What are the three most important challenges described in this conversation?*
- *Which parts of the conversation suggest areas that need further exploration?*
- *What follow-up actions would be most useful based on what was discussed?*

AI can process the full conversation in a way that is difficult to do from memory alone. It can identify patterns that you might miss and highlight key moments. It suggests the next steps grounded in what the person said.

3. Turn Insight Into Action

The output of this process should lead to clear next steps, such as:

- follow-up questions
- resources to share
- meeting recommendations
- specific topics to explore further

Remember: the goal is not to analyze the conversation for its own sake but to continue it with *greater clarity*.

4. Close the Loop

When you prepare, focus, and consistently capture, each conversation improves the next. You arrive better prepared, understand key concepts more quickly, and follow up with greater precision. Over time, the Conversation Loop turns individual conversations into a system that *not* only achieves successful results—it *compounds* them.

CHAPTER 10
TEACHING THE SYSTEM

Priya's next message arrived late on a Thursday afternoon.

Meg, I've been experimenting with the ideas you suggested. I tried them with my leadership team this week. The conversation changed immediately. But I realized something else. I understand the questions, but I'm really curious about how you prepare for these conversations so quickly. Would you be willing to show me how you do it?"

For the first time since Meg had started building her AI system, she was being asked not about human resource issues, but about the *process* she used to prepare for key interactions. She replied with a short message: *Of course. Let's walk through it together.* They scheduled another Zoom call.

Prepping for the Meeting About Preparation

As usual, Meg opened Priya's chat a few minutes before their meeting was to begin. There, the first two conversations with Priya were already documented: their initial meeting at the Bay Area leadership conference and the Zoom meeting from two weeks earlier. Each conversation had notes, transcripts, and follow-up reminders. Priya's chat also contained Meg's other research, including podcast and presentation transcripts, as well as company research. When Priya joined the call, Meg shared her screen. "You wanted to see how this works," she said. "Here it is." Priya nodded. "Yes. Because it looks effortless when you do it." Meg gave Priya a few minutes to absorb what she was seeing on the screen. At the top was Priya's name and role. Below that were sections Meg had built over time: conversation notes, transcripts, articles, and follow-up reminders—information that was both relevant and well organized. On the right side were prompts Meg had given the AI, and the AI's answers. Questions generated from earlier conversations. Observations from transcripts. Ideas for future discussions. Priya leaned closer to her screen. She looked intrigued. "You keep everything in one place." "Yes," Meg said. "So the conversation never starts from zero," Priya noted, seemingly enthralled by the depth of the file that Meg had built on her.

Priya studied the page. "I see you're using AI to help you. I tried doing something like this last week." Meg looked up. They had even more in common than she had thought. "How did it go?" Priya laughed. "Not well. Similar to what you've done here, I asked AI to generate questions for a meeting I had. The questions sounded smart, but the conversation felt awkward." Meg tilted her head slightly. "What kind of questions did it give you?" Priya consulted her notes. "They were things like, 'What leadership strategies are you using to scale your teams?'"

Meg smiled. "That's what I call a 'resume question,'" she said. Priya frowned. "What do you mean?" "A 'resume question' asks someone to explain their leadership philosophy," Meg said. "People answer it the same way they describe themselves in a profile or on a conference panel. But their answers don't necessarily provide context about their current challenges and problems." Priya considered that. "So, the question points backward." Meg nodded. "Exactly! Toward how someone describes themselves. The real conversation usually begins when you ask about what is happening *right now*." Priya leaned back in her chair. "I asked the wrong kind of question." "Not wrong," Meg said. "Just incomplete." Meg used her mouse to highlight a passage from the transcript of their previous meeting. "See how you told me something important here?" Priya read the sentence aloud. "Our managers delay releases because they keep solving technical problems themselves." "Right," Meg said. "That was an important insight for me because it showed me your real issue. And that's where the conversation starts. Now, look at this." Meg typed another prompt into the chat.

Based on this specific problem, what activity questions could help Priya explore this issue with her leadership team?

The system generated several suggestions:

- *When engineers become managers, what makes it hardest for them to step away from coding?*
- *What signals tell you a manager is still solving problems personally instead of leading the team?*
- *How do deadlines change when managers stop doing technical work themselves?*

Priya's eyes widened. "Wow! "You see the difference?" Meg asked. Priya nodded reflectively. "These questions aren't about the person. They're about the *problem*." Priya leaned back in her chair. "That's where I got stuck." They spent the next twenty minutes reviewing how the system worked. Meg showed how each conversation built on the previous one—from their first conference exchange, through all the research she had done, to the Zoom meeting and the subsequent transcript analysis. Each step had produced new questions for the next conversation. Priya pointed to the transcript section. "You're not physically taking notes during the meeting?" Meg shook her head. "No. I record the meeting." Priya raised an eyebrow. "The AI isn't giving you questions to ask in real time?" "No," Meg said. "I use the AI afterward to help analyze the meeting and prepare for the next one." "Instead, you listen to the person." The way Priya said it made it sound like a revelation. "I think I understand the preparation now. But what surprises me most is this part," Priya continued, pointing at the transcripts. "It allows you to remember everything." Meg nodded. "That's the difference," Priya said with satisfaction. They brainstormed for another twenty minutes about how Priya could build her own AI prep tool to help her achieve her networking and human resources goals. Before ending the call, Priya asked, "You know what this reminds me of?" Meg waited. "A good engineering system." Meg smiled. "That's exactly what it is." Priya looked thoughtful. "Networking isn't really about meeting people." Meg raised an eyebrow. "What do you think it is?"

Priya glanced again at the screen filled with conversations. "It's about building a system that remembers your conversations." "That's part of it," Meg replied. Priya looked curious. "What's missing?" Meg paused for a moment and then said, "The system remembers the conversations. But the relationship grows in the parts the system can't do." Priya waited. "The listening," Meg said. "The curiosity. The moments when someone says something important, and you decide to explore it instead of moving on." "Ah, the system keeps the record," Priya said. Meg smiled. "And people build the relationship." After the call ended, Meg uploaded the transcripts to Priya's chat and typed some notes. After she closed her laptop, she mused that, for months now, she had been building the structure for herself. Now, Priya was starting to build her own system to suit her needs. Her system was multiplying. And that was incredibly satisfying.

Key Take away: Teaching the System

When Priya tried to use the system on her own, she made the mistake most beginners make, the same one that Meg herself had made. She asked AI to generate questions, but those questions weren't based on relevant information. The questions sounded intelligent, but conversations based on them felt awkward. Why? Because the questions focused on *identity* rather than the real problem in the conversation. This mistake had taught Meg—and now Priya—an important lesson. Systems may seem simple to design and implement, but it takes practice to apply them thoughtfully and use them for the right things.

Teach Others, Teach Yourself

Chapter 10 introduces a new stage: you begin teaching the system to someone else. The moment you explain it, some interesting things happen:

- Your own understanding becomes clearer.
- You notice where beginners get stuck.
- You see which steps matter most.
- And you begin to describe the process more precisely.

The fastest way to refine the system is to teach it to others.

Structure, Not Just Theory

When introducing the system to someone else, don't start with an explanation. Start with the structure in these steps:

1. Start with the project and the chats.
2. Show how you gather and organize information.
3. Show how earlier conversations generate better questions for the next conversation.
4. Show the transcripts and how they capture details of the meeting that might easily be forgotten or missed in notes.

Once someone sees the structure, the method becomes obvious.

The System Remembers

At first, Priya thought that Meg had an excellent memory that could recall even minute details of their previous interactions. But Meg corrected her: the system remembers. Transcripts capture the conversation. Notes preserve the important signals. AI tools can help identify patterns and follow-up questions. With the system holding the information, you are free to focus on listening.

The Network Multiplies

The most important moment in Chapter 10 occurs when Priya starts *building her own system*. This changes the dynamic: Meg's method stops being a personal, individualized tool. It morphs into something bigger, something *transferable*. When others begin building their own systems, something powerful happens: the network expands through shared practice. People begin having better conversations. Those conversations produce better follow-ups. Network quality improves for everyone. The system keeps the record, and the people build the relationship.

CHAPTER 11
SEEING THE PATTERN

Two weeks after the conversation with Priya, Meg evaluated the progress of her project. Her system had continued to grow, and now included dozens of projects, dozens of conversations, plus transcripts, notes, follow-up reminders, and questions layered on top of each other.

As she opened her main networking project, names filled the screen, each with a chat—engineers, product leaders, founders, students, and managers. Each name was associated with conversations that spanned time. Until now, Meg had used the system one conversation, one meeting at a time. Prepare. Focus. Capture. But the system had reached a different stage. There was enough history inside it to ask a larger question. She opened the AI assistant and typed.

Review all stored conversations across my networking projects. Identify patterns in the challenges people describe, the questions that produce the strongest discussions, and the themes that repeat across industries.

The system paused. Then the analysis appeared.

Stop Solving, Start Guiding

The first pattern surprised her. Across nearly every conversation, according to the AI, people described the same sorts of transition-related growing pains: engineers becoming team leads and specialists becoming decision-makers. The real challenge was learning to stop solving problems personally and start guiding others. She had seen this problem before, but she had never realized how often it appeared.

Ask Activity Questions

The second pattern the AI found focused on questions. Certain questions consistently produced deeper discussions. Not identity questions. Not even resume questions. *Activity questions.* These were questions about decisions, trade-offs, and real situations. The system highlighted several examples.

- *What signals tell you a manager is still solving problems personally instead of leading the team?*
- *What changed in your team once you stopped answering every technical question yourself?*
- *What is the hardest part of letting someone struggle with a problem before stepping in?*

Meg recognized those questions. They had produced the longest conversations.

Picking Up Threads

The third pattern appeared in the follow-up notes. Whenever a second meeting occurred, the conversation almost always reopened with a detail from the previous discussion. This might be the continuation of a discussion about active problems or the revisiting of a previous decision. Whatever it was, the most productive conversations picked up the thread of a previous conversation. This worked both ways, but the take away from this for Meg was that people were more interested in continuing a previous conversation than starting a new one from scratch. This was valuable information.

Patterns

What other patterns would the system be able to find with more data? For months, Meg had been building her system one conversation at a time. Now the system was showing her things she could never have seen inside a single meeting, not even with the best analysis: patterns. The system not only remembered individual conversations. It revealed their overarching themes.

Building the Human Network

The next afternoon, Priya invited Meg to speak briefly with her team. "Nothing formal," Priya said. "They want to understand how the system works." Five people joined the call. After introductions, one of the managers asked the first question: "How do you prepare for conversations so quickly?" Meg shared her screen. At the center was a diagram. Three circles connected by arrows.

Prepare → Focus → Capture

"You prepare before the conversation," Meg explained. "You focus during the conversation, and capture what matters afterward. Then the cycle repeats." Another of Priya's managers studied the diagram. "So, preparation means research?" "Partly," Meg said. "But preparation also means reviewing the last conversation." She opened Priya's chat. The transcript from their earlier meeting appeared. Meg highlighted a sentence: *Our managers delay releases because they keep solving technical problems themselves.* "Priya said this in a previous meeting," Meg said. "This line was the thread that created the next conversation. Preparation means noticing moments like this and building questions from them." Another manager asked something more specific. "How do you notice those moments while someone is talking?" Meg thought for a moment. "You listen for friction." The group looked curious. Meg continued, "When people describe something that slows them down, frustrates them, or keeps repeating in their work, that is usually where the real conversation begins. You can hear that frustration in their tone or the words they use." The manager nodded. "So, the goal isn't to ask smart questions." Meg smiled. "The goal is to notice interesting problems."

Another question followed: "How do you avoid interrupting when someone is explaining something complex?" "You wait a bit longer than feels comfortable," she said. A few people laughed. "Most interruptions happen because we think we already understand the point," she continued. "If you let someone finish the full explanation, they often reveal the real problem in the last sentence." Several people were already scribbling notes. The conversation continued like this for nearly an hour. The questions gradually moved away from tools and AI and toward conversations. Questions like the following:

- "How do you decide which conversations deserve a follow-up?"
- "How do you continue a conversation months later without starting over?"
- "How do you remember the small details people mention?"

Meg answered based on what she had learned from creating and using her system.

- "You follow curiosity."
- "You capture the moment that mattered."
- "And you return to it later."

After the meeting ended, Priya and Meg remained on the call. "When I first started using the system, I thought it was about organizing conversations," Priya said. "But after seeing the analysis you ran," she continued, "I see something else." Meg tilted her head. "What do you see?" "The system shows where the same problems appear again and again," Priya answered. "It's like zooming out. Instead of seeing one conversation, you start seeing the landscape." Meg smiled. "That's a good way to describe it." Later that evening, Meg returned to her networking project, taking it all in once more: hundreds of questions over dozens of conversations, enabling her system to find unexpected patterns. A crucial realization hit her. She had not only been building a network; She had been building a *map*.

Key Take away: Seeing the Pattern

When you start using an AI networking system like this, you focus on individual conversations. Prepare before the conversation. Focus during the conversation. Capture what matters afterward. At first, each conversation feels separate. You prepare for one person, ask questions about one problem, and record one discussion. But over time, something changes. The system accumulates history. Conversations build on one another. Notes and transcripts accumulate across dozens of people and situations. At that point, a new capability becomes available: patterns.

Patterns Reveal What Matters

When Meg reviewed all her conversations, three themes emerged:

- challenges recurring across industries
- individual contributors becoming managers
- technical experts learning to lead instead of solving every problem themselves

These challenges appeared across different companies, roles, and contexts. But the underlying theme was the same. The system made this visible. With the help of her AI system, she also learned that activity questions consistently provoked deeper discussion. These questions addressed decisions, trade-offs, and real situations. Finally, she learned that the best follow-up conversations continued the thread of the previous conversation. This was the third pattern. And not only did Meg's AI system help her see that pattern, but it would be enormously useful for keeping track of those threads in the future.

Zooming Out

At first, the system helps you prepare for conversations. Later, it helps you understand them. When enough conversations accumulate, the system begins to reveal the landscape of problems people face, like these:

- different people struggling with similar transitions
- questions most likely to unlock insight

- opportunities to focus on next
- recurring themes, instead of isolated discussions

This is the moment the system becomes more than a networking tool. It becomes a map of conversations, problems, and meaningful discussions. Once you see the whole map, each new conversation becomes easier to navigate.

CONCLUSION WHERE CONVERSATIONS LEAD

Six months after the turning point conversation with Priya's team, Meg noticed something different about her calendar. Yes, it was more populated with meetings than it used to be. But there was something else: she wasn't having to go out and find people. They were coming to her. Requests for meetings filled her inbox, starting with messages like these:

- *Daniel suggested I speak with you.*
- *Priya mentioned the system you showed our team.*
- *Mark mentioned your name when I asked him about employee wellness programs.*
- *I read the notes from your conversation with one of our managers.*

That was progress. Months ago, Meg was having to hit the business networking circuit to find leads for her consulting business. Now, with so many leads coming in, she didn't have time to attend many events.

Another Coffee with Dan

Later that week, Meg met Daniel Brooks for coffee. They hadn't spoken in person for several months. Daniel arrived and sat down with a smile. "I think you'll want to see this," he said. He opened his laptop and turned it so Meg could see it. "This is the leadership program we talked about." Months earlier, Dan had described the difficulty his engineers faced when they moved into leadership roles. They had spent nearly an hour exploring the problem. Now the idea had become a pilot program inside Dan's company. "We tested it with three teams," Dan said. "The results were better than we expected." Meg studied the screen. "What changed?" Dan pointed to the first section. "We stopped asking people about leadership philosophy." Meg listened as he explained the adjustments they had made. The new questions focused on real situations managers faced every day. "What do managers do on Monday morning?" Dan said. "Who do they talk to when a project goes wrong? What signals tell them a team is starting to drift?"

When he finished, Daniel leaned back, looking proud of the system he had built with Meg's help. He seemed more relaxed than he had in their previous meetings. "We built something from that first conversation." "Amazing," Meg said. "Congratulations." It was an outcome neither could have envisioned when they first met. Dan closed the laptop. "You know what surprised me most? You never tried to sell anything." Meg smiled. "I was trying to wrap my mind around the problem." "I think that's why this worked so well," Dan said. "Your thoughtfulness and acknowledgment that you didn't have all the right answers right at that moment. You really listened to me. They spent the rest of the meeting discussing what the next phase of Dan's pilot program might look like:

- more teams
- more managers
- more conversations

A few days later, Meg sent a short proposal outlining how she could help expand the program. Dan reviewed it, signed the agreement, and returned it with the initial payment. Meg looked at the number for a moment. It was a substantial retainer. Larger than any single engagement she had signed since launching her consulting practice. More importantly, it confirmed that the system she had built to guide her conversations, organize her research, and track relationships was working. AI had not replaced the human side of networking. It had strengthened it. Pre-meeting prep, organizing ideas and insights, and follow-up reminders—all of these worked seamlessly thanks to the AI preparation partner Meg had created. It freed Meg to focus on the parts of the meeting only a human could do: listening, understanding, and building trust. Later that afternoon, Meg sat in her office and opened the system she had been building for months: projects, chats, conversation transcripts, activity questions, and follow-up reminders. The same structure that once helped her prepare for a single meeting now supported dozens of relationships and active clients. She opened Daniel's chat. Their first conversation. The notes from their second meeting. The transcript in which Daniel described the challenge his engineers faced when they became managers. Now Meg added a new entry:

Leadership program expansion. Daniel's team.
Next step: workshop design discussion.

Meg closed the laptop and checked the calendar for next week. It was full: more conversations, more introductions, more problems waiting to be explored. Some conversations wouldn't go anywhere, but most would at least expand her insight. And sometimes one good conversation would turn into something much larger.

Now, with the right tools, those conversations could grow even further. Meg was building her consulting practice with growing confidence, one relationship, one conversation, and one client at a time.

FINAL REFLECTION: WHAT HAPPENS WHEN YOU STAY IN THE CONVERSATION

Opportunities rarely appear because you chase them. They appear because you stay in the conversation. You listen. You ask better questions. You remember what matters. Over time, the conversations connect. Patterns emerge. Ideas grow. Relationships deepen. Pushing for outcomes doesn't help; you must stay engaged and curious long enough for them to emerge.

How to Begin an AI System

If you want to begin, start simply:

1. Create a project and start a new chat for every new prospect.

2. Record conversations. Collect relevant research and notes.
3. Use an AI assistant to help you see patterns and ask better questions.
4. Find the threads that will propel you to the next conversation, the next event, the next lead.
5. Repeat.

In the age of artificial intelligence, the professionals who succeed will not be those who use AI to replace human interaction, but those who use it to help deepen relationships.

Good luck.

ABOUT THE AUTHOR

Doron Noyman is an author, keynote speaker, and networking expert with more than three decades of experience in business development and community leadership in Silicon Valley. Over the course of his career, he has attended more than 2,500 networking events and built a reputation for turning conversations into lasting professional relationships, partnerships, and business opportunities. He previously served as Chair of the Silicon Valley Chamber of Commerce and currently serves as Chair of the Sunday Friends Foundation. His work focuses on helping professionals turn conversations and relationships into opportunities for growth. He holds a BS in Economics from Virginia Commonwealth University and an MSt in Law from the University of Southern California Gould School of Law.

WHO SHOULD READ THIS BOOK NEXT

If this book helped you, there is a good chance someone around you needs it even more. Networking is one of the few skills that influences success across most careers. Yet very few people are ever taught how to do it well. You probably know someone right now who struggles with this, maybe someone early in their career who feels invisible in meetings. Or someone starting a small business who knows their craft but does not know how to build relationships. Perhaps you know someone talented who works hard but does not know how to open doors. Someone who avoids networking events because they feel awkward or anxious. If one person came to mind while you were reading this, consider sharing the book with them. Many careers change not because of a single big opportunity, but because someone helped them learn to build meaningful connections. Sometimes, the most valuable thing you can do for someone is to show them where to start.

I WOULD LOVE TO HEAR FROM YOU!

Thank you so much for reading this book. It means the world to me. If you found it helpful, inspiring, or just enjoyable, would you take a moment to leave an honest review? Your feedback not only helps others but also keeps me motivated to create more valuable content for you.

Here's how you can leave a review:

Visit your Amazon Orders page, find this book, and click "Write a Product Review."

Thank you for your support!

—Doron

doronnoyman.com

Also by Doron Noyman

Available on Amazon:

From Invisible to Valuable: How to Network Without Anxiety: 5 Steps to Open Doors and Boost Your Career

From Invisible to Valuable: Skills for Career Success: Stories and Practical Steps for Teens and Young Adults to Master the Skills No One Teaches You but Everyone Expects You to Know

Made in the USA
Monee, IL
04 November 2026

67380769R00066